WHICH NEW ZEALAND SPIDER?

Including their Eight-Legged Cousins: the Harvestmen, False Scorpions, Mites, Ticks and Sea Spiders

ANDREW CROWE

PENGUIN BOOKS

Introduction

Insects [Class: Insecta]	**Spiders** [Class: Arachnida. Order: Araneae]	**Other Arachnids** [Class: Arachnida]
• **SIX** legs • Body divided into **THREE** parts • Antennae	• **EIGHT** legs • Body divided into **TWO** parts • **NO** antennae	• **EIGHT** legs • Body **UNDIVIDED** • **NO** antennae

The palps or 'feelers' of a male spider tend to be much larger than those of females.

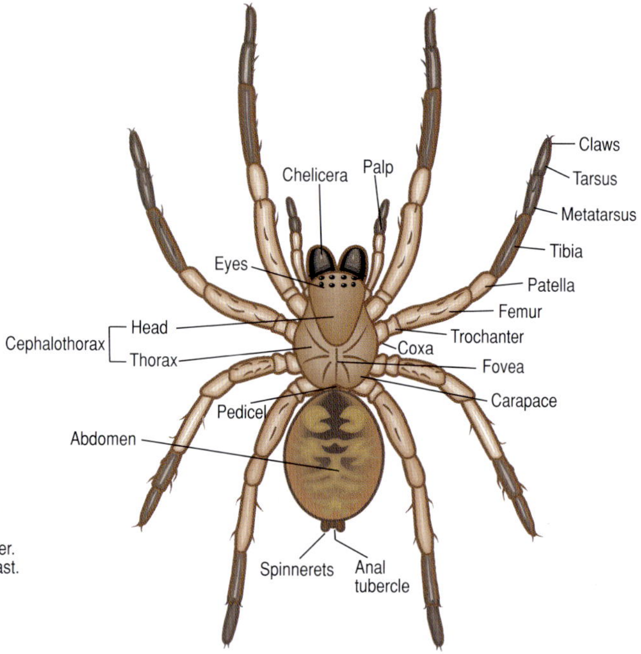

To use this book you need know no technical terms, so don't be put off by this technical body plan of a spider. It is only here for the spider enthusiast.

Spiders belong to a group of eight-legged invertebrates† known as arachnids (after Arachne, a skilled weaver in Greek mythology). The arachnid group also includes **scorpions** (not found in New Zealand), **false scorpions**, **harvestmen**, **mites**, **ticks** and **sea spiders**. Arachnids have been around for even longer than insects – for at least 360 million years. They are surprisingly adaptable, being found in a variety of habitats from the hottest deserts to the frozen slopes of Mt Everest at 7000 metres; from the darkest caves to deep beneath the sea.

Spiders generally hunt live creatures, mostly insects, often subduing them first by injecting them with poison. Many weave silken webs to trap their prey, and most have eight eyes. The number of spider species in New Zealand is estimated to be about 2500 (1300 of them so far named), all but approximately 40 of them native. Worldwide, the number currently known stands at around 39,500. (June 2006)

Spider life cycle: eggs hatch as little spiders, meaning that they do not undergo metamorphosis – like, say, a butterfly does.

---

† A creature without a backbone.

How and Where to Find Spiders

Indoors / Outdoors The best-known spiders are those which commonly live in and around houses, for example: **daddy longlegs**, **grey house spiders** and **whitetailed spiders**. However, most spiders far prefer life in the forest, in flowers, on the beach, in the mountains or by streams.

By Day Most spiders only hunt at night; only a few hunt actively during the day. Try sweeping a net through long grass to find these day hunters (but do not disturb their webs more than necessary, as spiders usually reuse the same web). In the daytime, you'll find most web builders hiding near their webs in a special retreat, or dangling on the end of a silk thread. Often, you can bring the spider out of its retreat by touching a vibrating tuning fork against its web, using a little paintbrush to block off the spider's retreat so that you can get a good look. By day, many night spiders can be found sheltering, hidden under bark, logs, stones, leaf litter or at the base of shrubs.

By Night A small headlamp is useful. Not only does this leave your hands free, but by having the light at the same level as your eyes, it will reflect off the back of the spiders' eyes, showing up as tiny green dots, which can be seen well before the spider itself. Soon after dark, you'll find the web builders starting to repair their webs, while hunting spiders are running about, resting on tree trunks, on leaves, or on the forest floor.

There's Life in the Leaf Litter

With so many spider species in New Zealand, you might wonder where they all are. Most are small and live in forest, so go unnoticed. In the forest leaf litter, their numbers generally peak from December to April. In December, for example, about 850,000 spiders and **harvestmen** are estimated to emerge from one hectare of rātā-rimu forest. And the number of tiny **mites** born in the same leaf litter is reckoned to be about 250 times more. Scientists who have studied the diets of native forest birds (such as tits, fantails, grey warblers, whiteheads, riflemen and hedge sparrows), have discovered that spiders are one of their most important foods, particularly in the warmer months – between October and February.

If You Have Trouble Finding a Spider in Winter, it is Because . . .

Most of the common spiders seen around homes and gardens spend this time quietly, with little action. In summer, they grow rapidly to reach maturity and mate. The life span of most species is often no more than a year. Those that live longer generally survive for less than three years. **Trapdoor spiders** (the exception) can live up to 20 years. During summer, when spiders are actively growing, their old skin soon becomes too tight for them (like a suit of armour) and they need to shed it. At this time of year, you will often come across the discarded empty skins of larger spiders lying around (see illustration on page 50).

Flying Spiders (Ballooning)

Unlike many insects, spiders have no wings and cannot fly – at least not in the usual way. Nonetheless, the young of some newly hatched spiders, and the adults of other smaller spiders, like **money spiders** [Linyphiidae], can go 'ballooning'. To do this, they pay out threads of silk to catch the wind. (Some small caterpillars get around like this too.) Carried on rising currents of air, some spiders have been known to reach heights of 6000 metres or more, where they have been collected by passing aircraft. To qualify for this kind of transport, the spider must weigh less than about 25 milligrams. Floating gracefully through the air, these tiny spiders are travelling rather like a hot air balloon, which is why this style of spider travel is commonly known as ballooning. Most of our Australian **orbweb spiders** have arrived here in this way. Although spiders will happily balloon at any time of year (unless the weather is freezing), the best time to see them in action is when a gentle breeze is blowing during late summer and autumn. Keep an eye out then for streaming lines of silk glistening in the sunshine. Sometimes, whole fields can be covered in a delicate layer of spider gossamer; it can be quite a sight.

The breeze will sometimes take such spiders out over the sea. On one occasion ballooning spiders were seen landing on a ship in the Pacific, off Hawai'i, 1600 kilometres from the nearest land (where they rested a while before taking off again). But, risky though this means of travel is for the individual spider, it does mean that spiders are, in some cases, better than flying insects at finding new places to live. For example, in 1883, when the Krakatoa volcano erupted, it was nine months before the first animal life was seen there – and yes, you guessed it: it was a tiny spider!

So, although spiders have no wings, many certainly do take to the air. Indeed, one Australian jumping spider (the **flying spider**, *Maratus volans*) has developed an interesting variation on this theme. When leaping, the adult spider uses wing-like extensions on the side of its body to help it glide.

Silkmakers, Birds' Nests & Bullet-Proof Vests

Not all spiders build a web-like snare to catch their prey, yet all do use silk to hold and protect their eggs. The innermost layer of their egg sac is loosely spun like coarse cotton wool to protect the eggs against shock. The next layer is closely spun to keep out the rain. A tough outside coating then protects the eggs against being eaten. In some cases, this coating is mixed with debris to also provide camouflage (see page 10).

Spiders produce this silk by squirting a fine jet of protein syrup out their back end through little nozzles (called spinnerets). As this comes out, it hardens into silk. Up to six kinds of 'syrup' can be produced by one kind of spider, each one of a different strength or stickiness for a different purpose. This partly explains why spiders don't end up trapping themselves, their mates or babies in their own webs. (Many also oil their legs for the same reason.) The engineering involved in making these webs

– especially the cartwheel-style orbwebs – is so sophisticated that whole books have been written about it.

Spider silk is special. For a given thickness, it is stronger than steel, twice as elastic as nylon and more difficult to break than rubber. Indeed, some of the biggest and strongest spider webs can catch small birds and bats. The natives of New Guinea put this feature to good use by using spiders' webs as fishing nets. They offer a hoop of supple bamboo to a large orbweb spider, letting it spin a web across it about two metres in diameter. By dipping the hooped web net into the water, they are able to catch fish up to about half a kilogram in weight. By adding further webs over the top as reinforcement, they can increase the size of their catch to fish that weigh up to 1.35 kilograms.

Spider silk has also been used to make clothes – although it does take a lot of spiders to make the silk and a lot of flies to feed them. The clothes are also very thin. Indeed, even with seven layers on, you still need to remember your underwear. Spider silk has also been used for the cross hairs inside microscopes and telescopic gun sights. The US Defense Department and the Indian Defence Research Development Organisation have taken the use of spider silk further and are experimenting with it for making bulletproof vests. Why? The properties of spider silk are so far unmatched by man, for not only is this silk incredibly strong and elastic for its weight, but it is uncannily good at absorbing shock. This means that it is rare for a fly to bounce off a web and yet equally rare for it to break through it. The same is true, it seems, of bullets. Many birds also find spider silk useful as nest material, but spiders are good enough recyclers themselves; when finished with their webs, they often eat them.

Spider Venom

Spiders have a wicked way of eating. They feed by injecting poison and sucking on their prey. Indeed, with the exception of one family of **orbweb spiders** [Uloboridae] which have no venom glands, all spiders are venomous. However, just 1% of the world's spider species are able to inflict humans with a significant bite. Even those larger spiders which can bite people in self defence, are generally shy. In New Zealand, only the **katipō** (and much rarer **Australian redback**) can truly be considered poisonous – although the bites of a few others can be painful. The same, however, is not true in many other countries. The venom of one North American spider, the **black widow spider** (*Latrodectus mactans*), is 10–15 times more potent than rattlesnake venom, but – even then – the quantity of poison injected by the spider is so tiny that an untreated victim has a three to four times greater chance of recovery. So it has to be said that spiders are far less dangerous than many people believe and also that they are often blamed unfairly for bites quietly inflicted by other skin-piercing creatures such as mosquitoes and fleas. Life is not so safe, however, for male spiders. If they are to approach the female and avoid being eaten, they must approach the female spider very, very carefully, using a variety of codes and signals to soothe her into a passive mood.

Arachnophobia: An Irrational Fear of Spiders

In New Zealand no human deaths have been attributed to spider bites for many years, so it has to be said that in this country the fear of spiders is largely irrational. Compared to the risk of an allergic reaction to a bee or wasp sting (or the danger of traffic on the road), New Zealand spiders are relatively harmless. Yet many people are still scared of them. So why is this? Around the world, a fear of nature takes different forms. Most English children asked in the 1950s which animal they disliked the most chose the snake. In 1988, most voted for the rat, while most children in the US in the early 1980s picked the cockroach. New Zealanders surveyed today would probably pick the **whitetailed spider** – but to find the facts of what we really know about this creature, see pages 46–47 and 60–61.

Spiders on the Menu

As food for humans, spiders may not be as popular as snails, caterpillars and beetle grubs, but they are nevertheless enjoyed in many countries. In Cambodia and Laos, for example, you can purchase the larger and hairier ones barbecued on a stick. Some Chinese claim you can add ten years to your life by including spiders in your diet. Spiders are relished too by the native people of South America, southern Africa and Australia. One of the most popular edible spiders is the tropical **giant orb-weaver**, *Nephila* (page 18). 'Pick it off its web,' say the locals, 'bite off the large abdomen and eat it raw.' It tastes mild, apparently, like raw potato mixed with lettuce. (I haven't tried it.)

A tasty meal is also provided by the world's biggest spider, a furry creature, one of the South American tarantulas, the **goliath bird-eating spider** (*Theraphosa blondi*) of Brazil, Venezuela and Guyana. Laid out alive, it is big enough to cover your dinner plate. Ordinarily, the spider is rather shy and hides in a long, silk-lined earthen burrow from which it comes out at night to hunt. With a body larger than a tennis ball, the spidery meal it provides is substantial, but you must take care to remove the irritating hairs first. The locals put aside the spider's fangs (which are about 2 cm long) for use later, having discovered that these make excellent toothpicks.

To try this, however, you'll have to go to South America; they are not found in New Zealand.

To Get Started with Identifying Spiders

It is helpful to know that spiders can be conveniently grouped according to how they hunt and that those which use a web to catch their prey can in turn be grouped according to the design of the web they weave.

QuickFind Key to Spiders

Weave **WEBS** to catch their prey (active mostly at **NIGHT**)

	Go to page	
ORBWEB SPIDERS (master weavers)	9	ORBWEB
COBWEB SPIDERS	19	COBWEB
LADDERWEB SPIDERS	24	LADDERWEB
LINEWEB SPIDERS	25	LINEWEB
SHEETWEB SPIDERS (trampoline-style webs)	26	SHEETWEB
TUNNELWEB and **TRAPDOOR SPIDERS**	28	TUNNELWEB

Use **NO WEB** to catch their prey

SIT-AND-WAIT SPIDERS (**SUN-LOVING** spiders that sit still, waiting to leap on their prey)	32	DAY HUNTERS
ROAMING DAYTIME HUNTERS (**SUN-LOVING** spiders that wander freely in search of prey)	35	DAY HUNTERS
ROAMING NIGHT-TIME HUNTERS (Wandering spiders that normally remain hidden during the day)	44	NIGHT HUNTERS
ANCIENT CAVE SPIDERS (Spiders that spend their entire lives in darkness underground)	52	NIGHT HUNTERS
OTHER ARACHNIDS (Have eight legs but no narrow 'waist') (**harvestmen**, **mites**, **ticks**, **false scorpions** and **sea spiders**)	53	OTHER

Troubleshooting

If the creature you wish to identify has only **six legs**, it is probably not an arachnid at all but an insect. On the other hand, if it has **more** than eight legs, it is probably another kind of invertebrate such as a slater, hopper, centipede or millipede. (For help with these, try *Which New Zealand Insect?* or *The Life-Size Guide to Insects and other Land Invertebrates of New Zealand*.)

WEB WEAVERS

In this section of the book, you'll find the weavers of the spider world – spiders that construct a sticky web, patiently waiting on the web, or right next to it, for their prey to get trapped.

This stay-at-home habit makes these spiders easier to find than the roaming spiders (pages **34–51**), which do not build webs. A simple look at a web will often tell you which family the spider belongs to, for each group constructs its trap in a particular design.

Such web weavers can be divided into:

MASTER WEAVERS – that is, ORBWEB spiders, which produce masterpieces like the pair opposite, and

SPACEWEB SPIDERS or APPRENTICE WEAVERS whose webs are much less neat and tidy (pages **19–31**).

Where no magnification is indicated, photos are shown life-size.

Orbweb Spiders [Superfamily: Araneoidea. Families: Araneidae, Tetragnathidae, Nephilidae, etc]

Two typical examples of cartwheel-style **orbwebs** (shown here at half life-size).

Female spiders of this group generally construct neat cartwheel webs, using remarkably sophisticated engineering to create complex angles, tensions and connections. The precise design of these webs varies from species to species. However, the procedure is not automatic. As the spider builds, she continually tests and adjusts the loadings and tensions. Indeed, if a spider which has finished building its own web is transferred to another unfinished web, it investigates what the new web lacks and completes the job. Although these spiders are clearly using both wind and gravity in the first stages of laying out these webs, they are also capable of coping in a weightless environment. On 28 July 1973, NASA launched two orbweb spiders into outer space to see how they would manage with building a web on *Skylab*. For the first few weeks both spiders had trouble, but finally discovered entirely new techniques for building perfectly symmetrical webs. Like all web-builders, orbweb spiders are almost blind. In this family, the males are often much smaller than the females. Over 30 species are known in New Zealand. Worldwide, the number is over 4000.

Left: a native **mason wasp** (*Pison spinolae*), the female of which seeks out orbweb spiders to sting and fly back to her white mud-nest (centre). Here, she lays an egg in the body of each paralysed spider as fresh food for her newly hatched young. On the right, a more recent enemy of orbweb spiders is shown, a black-and-yellow-striped introduced **common wasp** (*Vespula* species).

Egg sac of the **garden orbweb spider**.

Above and below: the **garden orbweb spider** in typical daytime resting position – with 'knees' drawn up.

Where no magnification is indicated, photos are shown life-size.

Garden Orbweb Spider

Eriophora pustulosa (was *Araneus pustulosus*) [Family: Araneidae]

The young of this spider are believed to have originally floated here on air currents from Australia, making this now the commonest orbweb spider found in New Zealand. It comes in a range of sizes, patterns and colours (several examples illustrated left), however all have five large bumps on their back – arranged with two large ones at the top (or high point), and three small ones at the tail end. Colours range from shades of white, black, brown, green, yellow and even orange. It is only rarely found far inside forest, but is very common in gardens and on clotheslines, making most of the orbwebs seen there. The spider itself is usually active only at night, so this is the time to check out these webs. Explore with a torch to find the web owner resting upside down in the centre of her web or feeding. Many will laboriously rebuild their web each night. If not on her web, she will be waiting nearby with one leg touching a 'telegraph thread' which will signal to her when dinner has arrived. The shorter-lived males are much smaller. These do construct a little web of their own but, when mature, visit the female's web to mate with her; the males die soon afterwards. The garden orbweb spider is so hardy that it is even found in the Subantarctic Islands.

Bird-Dropping Spider

Celaenia species [Family: Araneidae]

Native Of all the orbweb spiders, this is the most unusual, for it breaks all the rules; it builds no orbweb. Its common name comes from the spider's strange, blobby shape and the fact that some species in this group are a dirty, blotchy white. They usually dangle upside down on a silk thread hanging below a leaf. This is an odd way for a spider to catch its food. Indeed, for a long time it was a puzzle to arachnologists how the bird-dropping spiders could catch anything. Yet they rarely went hungry. Moths just kept on flying right up to them and getting eaten. A close look at these 'kamikaze' moths showed that all were male – and in most cases, leafroller moths. It seems that the spider is able to lure these moths by producing a special scent which mimics the sex scent, or pheromone, normally produced only by the female leafroller moths to attract a mate.

Side view of the **bird-dropping spider**.

The **bird-dropping spider** specialises in capturing leafroller moths. ▶

Cryptic Orbweb Spider

Cryptaranea species [Family: Araneidae]

Native Common. Very variable and well-camouflaged. The top one is from Lake Alexandrina (central South Island); the bottom two from Mangawhai (Northland).

Where no magnification is indicated, photos are shown life-size.

White-Banded Orbweb Spider

Zealaranea crassa [Family: Araneidae]

Native Very common in pasture and gardens, often found on hedges, and among gorse and mānuka. Its markings are very variable, often showing a white band across the front of its abdomen. The spirals of its web are closely spaced.

Black-Spotted Orbweb Spider

Zealaranea species [Family: Araneidae]

Native On leaves, this spider achieves its camouflage by having two black patches on its back, as if mimicking the effect of holes in the leaf caused by insect damage. Unfortunately, too little is known about New Zealand orbweb spiders to give this one a more precise name.

The **green orbweb spider** is shown here in her typical night pose – hanging upside down in the centre of her web.

4x

6x

Green Orbweb Spider

Colaranea viriditas (was *Araneus*) [Family: Araneidae]

Native Common in forest and in gardens throughout New Zealand. A green leaf-like shield on the spider's back serves as camouflage, making it hard to find during the day. However, **mason wasps** (see page 9) still manage to spot them, for these spiders are commonly found paralysed in the wasps' nests, providing food for their young.

2x

Where no magnification is indicated, photos are shown life-size.

Two-Spined Spider

Poecilopachys australasiae [Family: Araneidae]

An Australian spider found in the upper North Island since the early 1970s. Now common down to about Wanganui. Seen in spring. The best place to find them is in citrus trees, where the female spider can be seen hiding under the leaves by day, emerging at night to construct her cartwheel-shaped web – which she recycles by eating again before dawn. Only she has the bright colours (which she can quickly change); the male (right) is duller, hairier, much smaller (2.5 mm) and lacks the distinctive spines of the female. Useful in the garden for catching large moths. Harmless to people. Her spindle-shaped egg sac is shown below, right. The two-spined spider belongs to a group of spiders known as bird-dropping spiders – so named because of their odd looks.

The male **two-spined spider** is much smaller and hairier than the female above it.

The mother spider is guarding her spindle-shaped egg sac.

Horizontal Orbweb Spider

Leucauge dromedaria (often misspelt as *Leucage*) [Family: Tetragnathidae]

An Australian spider which has become common in the North Island of New Zealand, in open forest and gardens. It builds a distinctive large horizontal web near the ground and hangs underneath this during the day. By hanging upside down, the silvery side of its body is cleverly disguised from below against the sky. When viewed from above, its dark underside is equally well camouflaged against the ground. If need be, it can use its long legs to run fast.

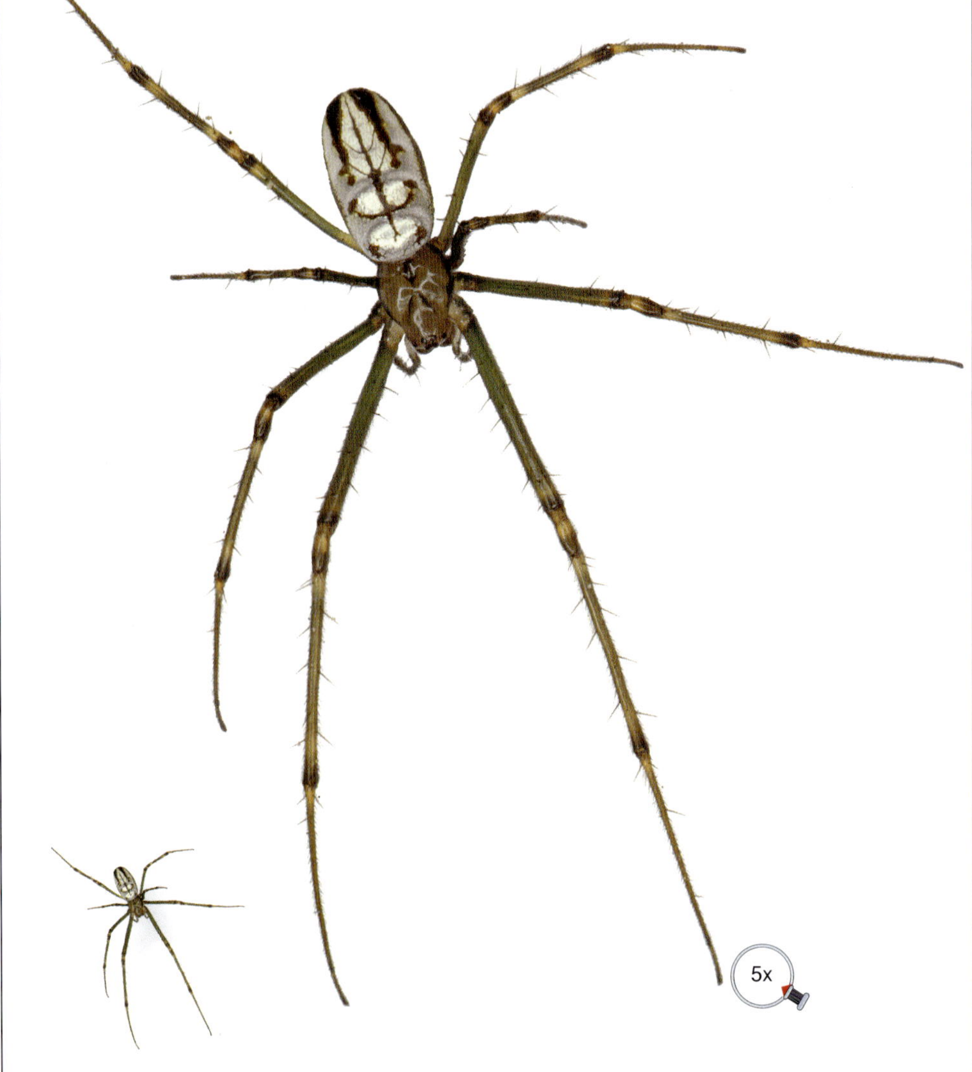

Where no magnification is indicated, photos are shown life-size.

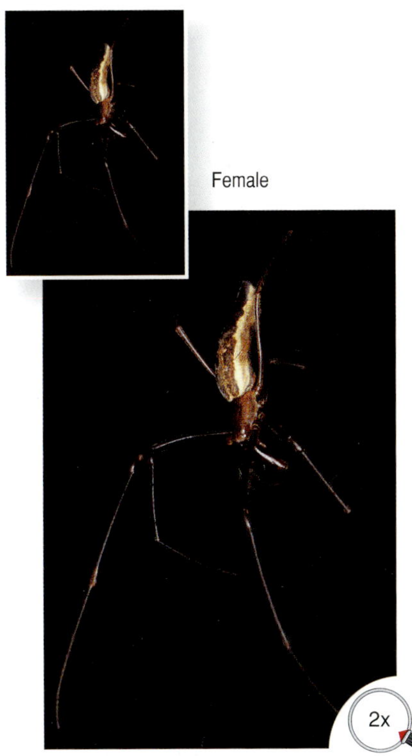

Female

Male

2x

Swamp Orbweb Spider

Tetragnatha species [Family: Tetragnathidae]

Native These and the family of orbweb spiders that they belong to are known as **four-jawed** or **big-jawed spiders**, all of which have very slender bodies and very long legs. Like most, this one lives in wet places, particularly near open swamps.

Stream Orbweb Spider

Tetragnatha species [Family: Tetragnathidae]

Native Clings underneath a horizontal web which it builds over streams, close to the water surface. When disturbed, it immediately drops into the water. To get a good look at it, you must approach it very, very slowly. The male is unusual (for a spider) in that it can make noises. Sadly, these are too quiet for us to be able to hear. To photograph this one from underneath its web, the back of my head is dipped into a cold forest stream near the top of the Coromandel Range.

2x

Orbweb Spiders (continued) [Family: Nephilidae]

The female **golden orbweb spider** is often seen on its web in full sunshine.

Golden Orbweb Spider

Nephila edulis [Family: Nephilidae]

From Australia and the Tropics. Sometimes seen in northern New Zealand. The female looks spectacular sitting in the centre of her web in the middle of the day with a body up to 24 mm long. (The male is just 6 mm long.) Over in Australia, the tiny young spiders put out a thread, letting the wind pick them up, bringing some to New Zealand most years. As the *edulis* part of this spider's name tells us, this spider is tasty to eat – for, in many tropical countries (including New Caledonia and Papua New Guinea), the larger female is roasted and eaten. The common name refers to the spider's three-metre wide, golden-coloured web, which is strong enough to catch small birds and bats. Indeed, researchers are currently experimenting with using these to stop bullets, designing bullet-proof vests made from them. There is a catch though: it takes 1300 webs to make one metre of cloth.

SPACEWEB SPIDERS

To distinguish the three-dimensional webs of the following web-builders from the perfectly neat, flat snares of orbwebs, they are called **spacewebs**.

Cobweb Spiders

[Family: Theridiidae]

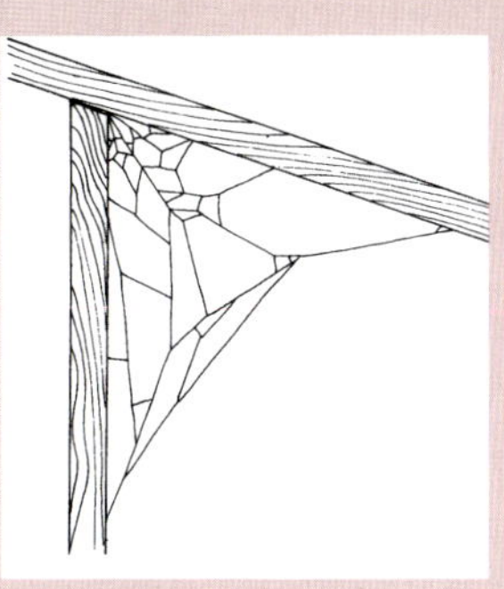

Cobweb spiders (or **comb-footed spiders**) build a strong, untidy web, and use combs of curved bristles on their back legs to throw silk over their victims. Most have a pea-like back end and are active only at night. While a few kinds live inside or under houses or in gardens, most are found well away from houses. Most are harmless to people, but the female of one overseas member of the family – found in many warm parts of the world – is famously poisonous: the **black widow spider** (*Latrodectus mactans*). Its local cousins, the native **katipō** (*Latrodectus katipo*) and **Australian redback** (*Latrodectus hasseltii*), are less dangerous. Over 20 species of cobweb spiders are known in New Zealand. The world figure of known species currently stands at over 2200.

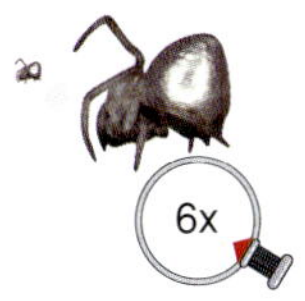

Dewdrop Spider

Argyrodes antipodianus (often misspelt as *antipodiana*)

6x

Australian. These tiny spiders – with a body length of just 2 mm – can and do build their own tiny webs, but they mostly rely on pinching their food from the backs of webs of other spiders. In the spider world, such creatures are known as 'kleptoparasites' or 'web pirates'. A favourite hangout is the webs of the common **garden orbweb spider** (pages 10–11). The adult dewdrop spider is so lightweight that it can employ the trick of paying out a silk thread in order to let itself be carried by the wind from the web of one host spider to the web of another, by 'ballooning' (see page 4). Although absent from the southern half of the South Island, this tiny spider is very common in the warmer regions of the country. The female can sometimes be seen guarding her cluster of tiny egg sacs, which are shaped like little toadstools. The spider's common name refers to the fact that, in strong sunlight, its gleaming, conical, back end can look like a drop of dew. It also looks rather like a speck of silver solder, inspiring the alternative name, **quicksilver spider**. Its own tiny web will catch food too, but is largely used instead as an escape route, a refuge, a snare or larder for storing its stolen goods. This escape route only works because the thief's webs are stickier than the web of its host.

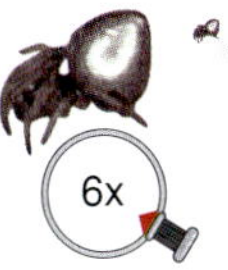

6x

Katipō Spider

Latrodectus katipo

Native The adult female New Zealand katipō has a poisonous bite. She is very shy, however, and when disturbed prefers to curl up into a ball. Only two people have ever died from the bite – both of them over 100 years ago. Nowadays, an anti-venom is available. She is about the size of a pea; the male is much smaller, lacks the red marking and does not bite. The katipō only lives near warm, sandy beaches and on sand dunes, especially around driftwood, cans, bottles and dune vegetation, where it builds its webs and eats mostly beetles and ground insects. It is found as far south as Greymouth on the west coast and Karitane on the east coast, but is no longer common. The name, katipō, means 'night stinger' or 'night biter'. The true katipō has a distinctive red stripe on its back. On beaches in the north of the North Island, a less common native **black katipō** (below) is also sometimes found, but this is not considered dangerous.

Female **katipō** with egg sac.

Black Katipō

Latrodectus katipo
(was considered as *L. atritus*)

Native This katipō has the characteristic, red hourglass underneath (illustrated at the top of the facing page), but lacks the red stripe on top. It is found only on North Island beaches, north of New Plymouth–Napier, usually in sandy dunes. To thrive, its eggs and young need more warmth than the red katipō. As of 2008, the black and red forms of katipō are considered to be variations of the same species, and thus likely to be equally poisonous.

Where no magnification is indicated, photos are shown life-size.

A characteristic red, hourglass design is found on the underside of the **Australian redback** (illustrated) and on the **katipō** (both species, opposite).

Australian Redback Spider

Latrodectus hasseltii (rather than *L. hasselti*)

Australian. Often intercepted in imported goods. In 1980 a few were noticed in Northland, Rotorua, New Plymouth and Wanaka, and several other places in the South Island, but only in Central Otago and New Plymouth has it become established. Here in New Zealand, this spider is unlikely to become common because of our cooler, damper climate. The Australian redback is slightly larger and more velvety than the native katipō, but experts can only tell them apart by examining their velvet-like hairs with a 10x hand lens. Being larger than the katipō, the redback can inject more venom, making it potentially more dangerous. It can also be more of a problem since it often lives near buildings. Up to an hour after the bite, a reddish lump appears with red streaks spreading out from it. Although most people recover with rest and a cold pack over the bite, an anti-venom for the bites from both spiders (redback and katipō) is available and this can be administered up to eight hours after the bite, relieving the symptoms within 24 hours.

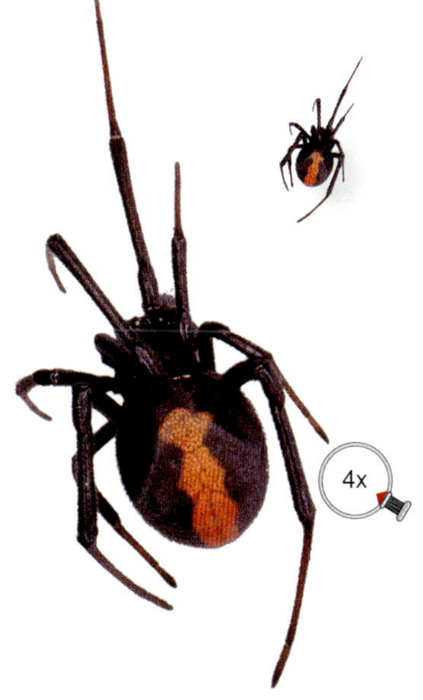

False Katipō Spider

Steatoda capensis

From South Africa and now widespread in New Zealand. This looks similar to the **katipō spider**, but has only a tiny, extremely faint, red marking at the tail end. Unlike the katipō, this spider is common around houses and in the garden, under bark, stones and flowerpots. It is also found near the beach, where it has taken over many of the sand dune areas in which the katipō normally lives. This is partly because it breeds more readily. Although the bite of the false katipō spider can be painful, it is not dangerous. Males can make faint noises by grinding a peg-and-ridge mechanism in their waist.

Square-Ended Cobweb Spider

Episinus species

Native This common, gold-and-brown spider comes out at night, usually in forest but sometimes in gardens. With a torch or headlamp you are most likely to find it among low-growing shrubs and ferns. Here, it catches its food by clinging upside down to a branchlet from just a few threads of silk. Easily recognised by the cut-off shape of its back end.

Where no magnification is indicated, photos are shown life-size.

New Zealand Cobweb Spider

Cryptachaea veruculata (was *Achaearanea*)

Native The webs of this spider are very common on the outside of houses and on fences. With overseas creatures and plants so often making their way to New Zealand, it is unusual to find a spider which has gone the other way: this one has been accidentally introduced from New Zealand to both England and Australia.

White Cobweb Spider

Parasteatoda tepidariorum (was *Theridion* or *Achaearanea*)

Worldwide. A recent arrival here, probably travelling in freight containers. Like the native **New Zealand cobweb spider** (above), it is common in damp places such as around light fittings on the outside of houses, but has so far not been seen in the South Island. Leaves untidy webs hanging from the corners of neglected ceilings of porches, basements and sheds, or in sheltered spots in the garden, resting in a retreat within the web. From late spring to late summer, light brown, pear-shaped egg sacs are seen hanging within the web. The female spider is illustrated; the male is much smaller. Also known as the **domestic spider**.

Ladderweb Spiders

[Family: Desidae]

The typical ladderweb (left, at one-third life-size) turns messy with age (right, life-size).

Spiders in this group generally build a zig-zag ladderweb which becomes messy with age, as shown in the pair of photos above. A particularly unusual species in New Zealand (not commonly seen) is the **marine spider** (*Desis marina*) which hides among seaweed on coastal rocks, spending much of its life under the sea catching small fish. All up, over 80 members of this family are known in New Zealand and 180 species worldwide.

Grey House Spider

Badumna longinqua (was *Ixeuticus martius*)

From Australia. Its webs are extremely common in sheds, on fences and car mirrors, on the outside of houses and around windows all over New Zealand. Not found in forest. In one corner of the web, a funnel can usually be seen, leading off to the spider's hiding place. The spider comes out to feed mostly at night. It is itself often eaten by the **whitetailed spider** (page 47) for which it appears to be a favourite food. A darker, less common relative is sometimes seen in the North Island, the **black house spider** (or **black window spider**, *Badumna insignis*).

Where no magnification is indicated, photos are shown life-size.

Lineweb Spiders

[Family: Pholcidae]

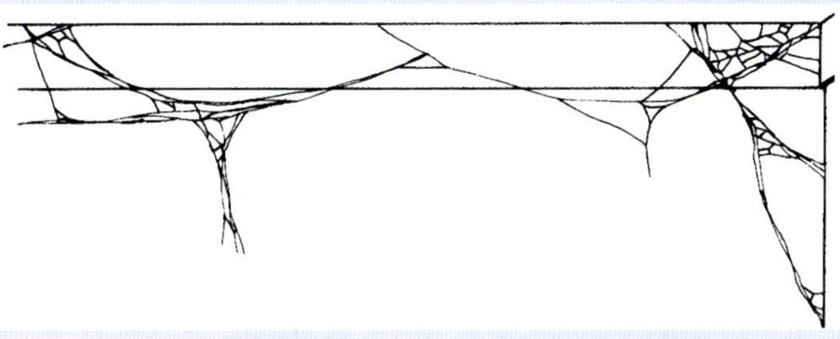

Lineweb spiders all have particularly long legs and weave loose and tangled webs, so are sometimes also known as **tangleweb spiders**. Often confused with **harvestmen** (see page 54). Three species are believed to live in New Zealand but just this one below is named. Worldwide, there are over 800 species.

LADDERWEB

LINEWEB

The spider's tiny body magnified

Daddy Longlegs Spider

Pholcus phalangioides

Introduced. Common inside houses throughout the world. It is common for books and websites to claim that the venom of this spider is one of the most poisonous spider venoms known. We are lucky, they say, because its fangs are too short to bite humans. However, there is no evidence to back up this claim. The spider's fangs are indeed short and it does have venom glands, but there is nothing especially poisonous about the venom. Indeed, the usual way this spider catches its prey is by simply wrapping it up very tightly with silk before sucking out the contents. Its sucking power is so great that the spider can empty a fly simply by sucking on the tip of one of the fly's legs. When something too big lands in its web, the spider whirls round and round, making itself almost invisible as it attempts to scare the intruder off. To see this, try touching a web. The female is often seen carrying her silken egg sac around with her. Can live up to three years. Also known as the **cellar spider**. (Confusingly, **daddy longlegs**, is a name which is also sometimes used for **harvestmen**, page 54 and for the completely different **crane flies** (**matua waeroa**).)

Sheetweb Spiders

[Families: Linyphiidae & Stiphidiidae]

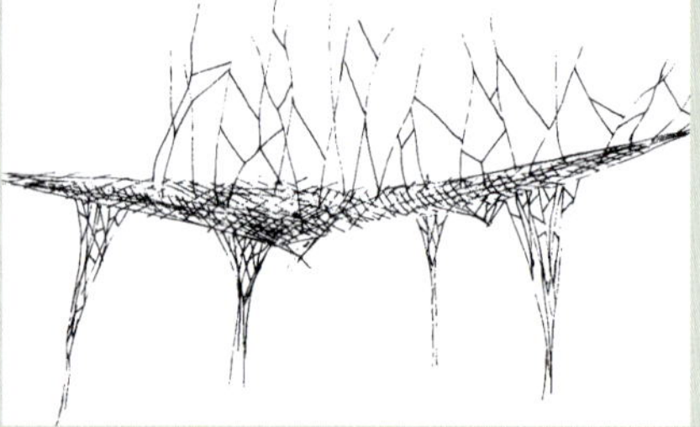

Web photo (left) one-sixth life-size.

Also known as **latticeweb spiders**. The horizontal, mist-like webs of these spiders are very common in the forest, often seen against the trunks of old trees, built like a trampoline with a network of special 'knock-down' threads above. At night, insects blunder into these threads, fall and get stuck in the 'trampoline'. By day, the spider rests in a short, silk-lined tunnel at the edge of the web, normally coming out only after dark. Active during the day is another, ant-sized, black kind – **money spiders** [Linyphiidae] – which produce the gossamer threads seen covering fields in autumn. Between these two sheetweb spider families, over 150 species are known in New Zealand, almost all native. Worldwide, the figure is over 4000.

Sombrero Spider

Stiphidion facetum (was *Amarara fera*) [Family: Stiphidiidae]

Australian. Although once believed to be native to New Zealand, this spider is now recognised as a recent immigrant from Australia. It is common in woodsheds, under the eaves of houses, under bridges, on clay banks and around rocky overhangs, where it weaves a pointed sheetweb shaped rather like an upside-down sombrero (a wide cone); hence the name. Beneath this the spider waits for insects to tumble in. Though common in the northern North Island, this spider has not yet been found in the South Island.

Where no magnification is indicated, photos are shown life-size.

A female **sheetweb spider**.

Sheetweb Spiders

Cambridgea species [Family: Stiphidiidae]

Native Throughout New Zealand, there are 29 known species, the largest of which can have a body up to 25 mm long. Very common in native forest and in gardens, where their distinctive trampoline-style webs are easily seen among the leaves of shrubs. The webs of the largest species from North Island forests (*Cambridgea foliata*) can be up to one metre across. During the day, these webs appear deserted; only at night is the spider seen hanging patiently under her web. It is not unusual however for the wandering males to come into houses, where they can fall into the bath and get trapped, unable to climb the slippery sides. Many of the males have hard ridges in their 'waist' which they grind together to make sounds (too quiet for us to hear). Since the pattern of ridges varies slightly from species to species, the sound from each kind is different.

Slender male with typically swollen palps ('feelers').

SHEETWEB

Underwood Sheetweb Spider

Haplinis species (was *Mynoglenes* species) [Family: Linyphiidae]

Native Common in forest and among mānuka scrub, where they build small sheetwebs beneath fallen wood and stones. In summer, they construct a small egg sac near the web – a low mound covered with white silk. I met this one among lettuce plants in my garden – not far from mānuka scrub.

Tunnelweb Spiders

[Family: Hexathelidae]

These large, hairy spiders all live in tunnel-like webs where they wait, ready to leap out and catch passing insects. The males, when looking for a mate, sometimes wander into houses. Unlike other spiders which bite by nipping their fangs together sideways, tunnelweb spiders (and their relatives, the **trapdoor spiders**) must press their downward-pointing, pick-axe fangs against the ground or a firm surface. If you spot this fang feature, you can be sure you have found a Mygalomorph. Mygalomorphs are large and ancient. A bite from a New Zealand tunnelweb can be painful, but does not usually lead to any serious effects. However, their east Australian cousins, **funnelweb spiders**, can indeed be dangerous. Worldwide, 85 species of tunnelweb spiders are known; 25 of them in New Zealand.

Banded Tunnelweb Spider

Hexathele hochstetteri

Native The web tunnels of these spiders are common on rocks, under stones and logs, and in little finger-sized holes in tree trunks. At night, the spider is waiting for insects to walk past the entrance. To get a good look at the spider, you will need to keep very still. Any movement and it will dash for cover inside the tunnel. Over most of the North Island, this is the commonest tunnelweb species, easily recognised by the chevron design on its back. Although its bite can be painful, it is not considered poisonous.

Where no magnification is indicated, photos are shown life-size.

New Zealand's heaviest spider, the **black tunnelweb** (above), has no trouble eating **garden snails** (below left).

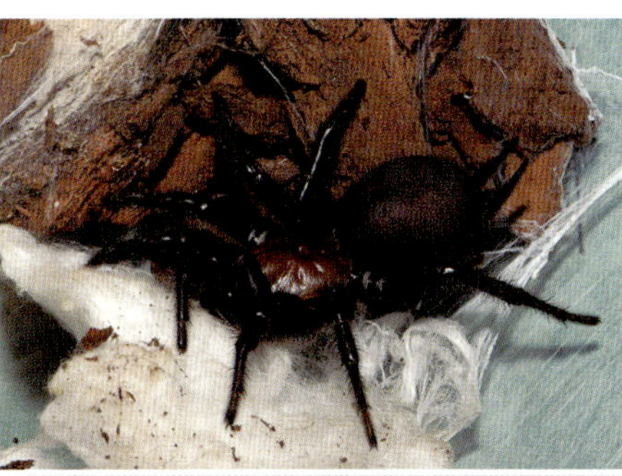

TUNNELWEB

Black Tunnelweb Spider

Porrhothele antipodiana

Native New Zealand's heaviest spider. Common in the South Island and around Wellington. In springtime, males wandering in search of a mate will often come into houses where they can easily find themselves trapped, put in a jar and brought into a museum for identification. As the museum arachnologist will point out, this species is darker coloured than the **banded tunnelweb** with less striking markings, its front section yellowish or orange and the legs coal black. Although the usual food of black tunnelweb spiders is millipedes, slaters and beetles, they also enjoy dining on one most unlikely creature, the garden snail. Now, snails not only have the ability to defend themselves with slime but they also have a shell to withdraw into. This doesn't deter the spider though, who quickly stabs the snail's body with its fangs, then holds on fast, letting the snail's gooey foam cover the front of its body while the snail tries to withdraw into its shell. Typically, the spider's escargot meal can last for 15 hours.

Trapdoor Spiders

[Family: Idiopidae]

Among the longest living of spiders, some survive for 20 years. Most are big. They live in tube-like tunnels up to 30 cm deep, which they excavate in the ground with their jaws and usually line with silk. These passages are spacious enough for the spider to turn around in. Although open grassland species cap their tunnel with a hinged, tight-fitting silk door, forest species build an open tunnel with no lid. The true trapdoor kind hide inside with the door slightly open. When they feel the vibration of a passing insect, they rush out to grab their prey and drag it down into the burrow. They do not truss their prey in silk, but simply grab it with their fangs. To protect themselves from hunting wasps, they are able to hold their lid down with surprising force. The tug on the door of one American trapdoor spider was measured and found to be 140 times the spider's own weight (equivalent to a human lifting a ten-tonne truck). Over 260 species known worldwide, of which over 40 are known in New Zealand, all native.

Above: spring-loaded spider trapdoor.

Left: one of the trapdoor spider's greatest enemies, a **large black hunting wasp** (*Priocnemis monachus*), which is strong enough to prize this lid open.

Trapdoor Spider

Cantuaria species (was *Misgolas*)

Native Most members of this family are found only in the South Island, where they hide from predators (and photographers) by excavating a burrow 25–30 cm deep. Using its fangs, the spider firms the walls of its tunnel with saliva before lining its new house with silk. They complete their retreat by constructing a wafer-like hinged lid of silk and soil, which fits snugly over the finger-sized entrance. The hinge is ingeniously spring-loaded with silk, so the spider can push this open easily from inside, letting it snap shut when released, leaving the entrance almost invisible. To be sure not to lock itself out, the female spider simply stays home. This spider has its enemies nonetheless, the greatest of which appears to be New Zealand's largest hunting wasp, the **large black hunting wasp** (above). In summer, these wasps go around collecting and paralysing trapdoor spiders and **tunnelweb spiders** (page 28) as food for their growing larvae (grubs). These wasps are often not only strong enough to pull the lid open, but some will even bite clean through the hinge of the trapdoor to get at the spider. It is not unusual for the wasp to proceed to take over the spider's burrow as its own, opening and closing the door as it comes and goes. Most trapdoor spider species with hinge-door tunnels live in open grassland, however a few are found in wet forest.

Where no magnification is indicated, photos are shown life-size.

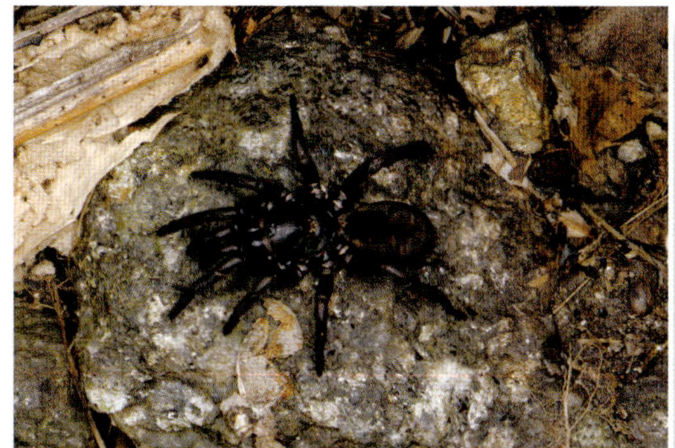

Although it is rare for **trapdoor spiders** to leave their tunnel, this one from Charming Creek, near Westport, was out roaming.

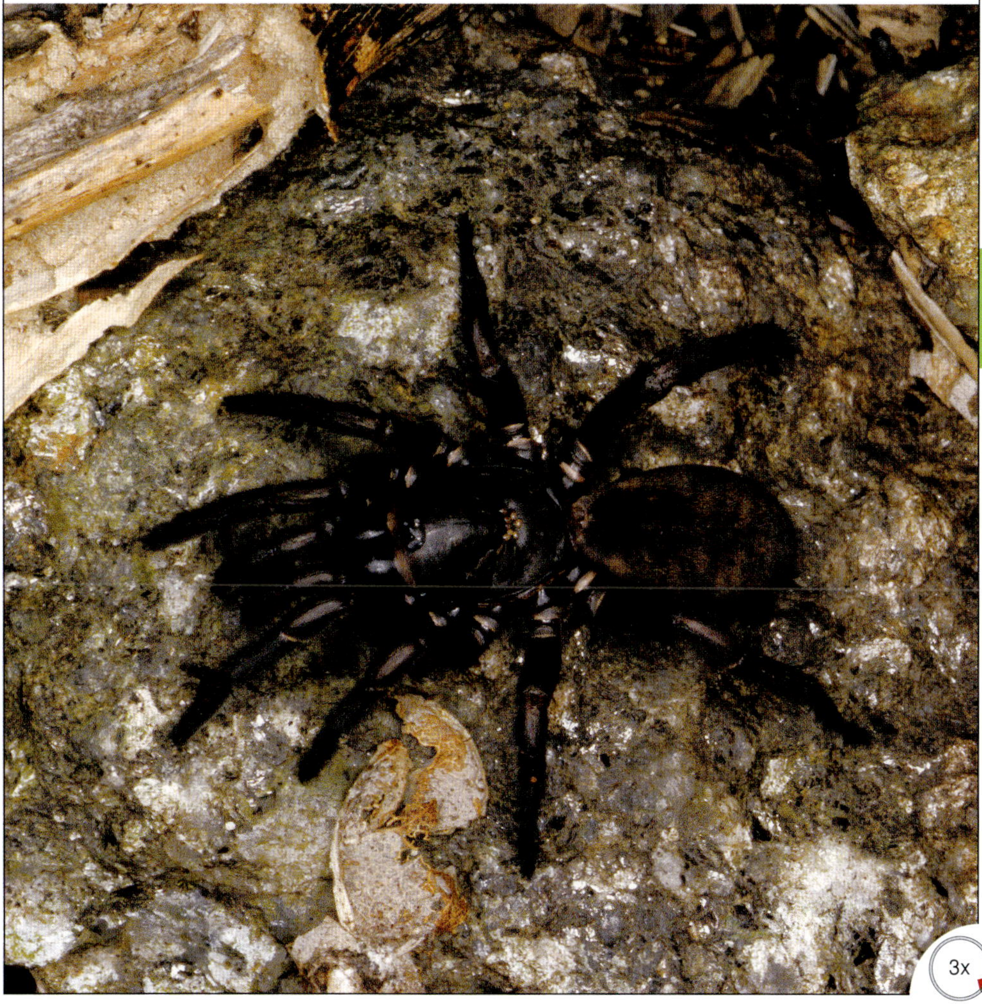

3x

Where no magnification is indicated, photos are shown life-size.

'SIT-AND-WAIT' SPIDERS

(Often active during the **DAY**)

Also known as **ambushers**. Like the roaming spiders in the next section, these spiders don't use webs to catch their prey. Instead, they keep very still, hidden by their own camouflage colouring, waiting for their food to come ambling by. They then pounce on their victims.

Crab Spiders

[Family: Thomisidae]

Rather than take the trouble to build a web, these little spiders usually hide inside flowers, waiting to catch a passing meal. They are named after the crab-like way they hold out their long front legs and the fact that they can scuttle sideways – also like crabs. Many are well-camouflaged according to the habitat they live in, often with bright colours to match the petal colours of flowers where many hide. Some can even change their colour to match their surroundings. One South American species (*Strophius nigricans*) has taken its disguise to extremes, having learnt to conceal itself inside ant colonies by killing one ant and carrying the dead ant around on its back, thrusting its head around as a challenge to other ants. Over 30 species are so far known in New Zealand; over 2000 worldwide.

4x

7x

Flower Spider

Diaea species

Native Like other members of this family, these spiders hold their long front legs out to the side in crab-like fashion, and can run forwards or scuttle sideways. During the day, they are often hiding in flowers, waiting to jump on passing flower-pollinating insects. Many can change colour to match their surroundings.

4x

4x

Where no magnification is indicated, photos are shown life-size.

4x

4x

2x

Square-Ended Crab Spider

Sidymella species

Native Active day or night, these spiders are well-camouflaged on tree trunks or leaves, or dangling on a single thread in the forest, where they wait to catch a passing meal. Some are smooth and some leather-like, but otherwise all are similar in that they mimic the colour and texture of the logs, tree trunks and leaf litter where they live. If disturbed, they will quickly abseil to the ground, cut their dragline and fall onto their backs. This is where their square back end comes in handy. After one or two minutes (or when danger has passed) the spider uses its long front legs to lever itself backwards over this square back end (in a sort of slow-motion backwards somersault) to land on its feet again. Twenty or so species are known from Australia and New Zealand.

4x

ROAMING HUNTERS

This section of the book is devoted to the wanderers of the spider world. Rather than staying in one place and constructing a sticky web to catch their prey, these spiders all have a roaming way of life, creeping around, running or jumping in search of their food.

These hunters can be loosely divided into three groups:

ROAMING DAYTIME HUNTERS (opposite)
These sun-loving spiders are the ones which are most often noticed. (Most of them have good eyesight.)

ROAMING NIGHT-TIME HUNTERS (page **44**)
Not until these night hunters get trapped inside the house will most people notice them. To get a better look, go outside at night with a torch.

ANCIENT CAVE SPIDERS (page **52**)
These 'living fossils' of the spider world spend their entire lives inside caves.

Where no magnification is indicated, photos are shown life-size.

Jumping Spiders

[Family: Salticidae]

These are the tigers of the spider world, hunting by day. They can often be seen leaping from place to place with a silk safety line attached. Apart from being remarkably good jumpers (covering up to 20 times their own length in one leap), they have great eyesight – indeed the best vision of all spiders. Two of their eight eyes stand out like old-fashioned car headlights and can zoom in like miniature telescopes. Their remaining six eyes are used for detecting movement. As they turn their head to peer at you with these 'telescope eyes' they can look almost human. They are among the shortest-living of spiders, with some fast-living tropical species surviving for just a few months. Both males and females are often seen dancing or posturing in front of each other (and will do the same if they see their own reflection in a mirror). Some species mimic ants. Two species of *Euophrys* live on Everest far above the highest plant life, at 7000 metres, thus sharing the world altitude record for a spider. Over 50 species of jumping spiders have so far been named in New Zealand (although there are thought to be at least 100 species here). Worldwide, over 5000 are so far known.

Blackheaded Jumping Spider

Trite planiceps

Native This spider's ebony-black head and large black front legs make it easy to recognise. Its long greenish abdomen is also marked with a characteristic yellow stripe. Often found hiding inside the rolled up leaves of flax and cabbage trees, but also common in summer on the inside walls of houses, where, like other jumping spiders, it hunts during the day. This one is able to leap 20–40 cm in one jump. Slowly move your finger near one, and it will often jump on and continue to jump the gaps between your open fingers. Peer at one closely and it will look back at you almost quizzically and may even jump up onto your nose, making a surprisingly soft landing. Although more common in the warmer North Island, these entertaining creatures are found throughout the country. If you are someone who finds yourself afraid of spiders, this one might help you get used to the idea that most are really quite harmless. Just remember to be gentle with them!

House Hopper Spider

Trite parvula & Hypoblemum albovittatum (now *Maratus griseus*)
(previously lumped together as *'Euophrys parvula'*)

There are now thought to be two very similar species in New Zealand, the first of these (*Trite parvula*) possibly native, the second one accidentally introduced from Queensland, Australia. Both are extremely common here, where they are often seen sunning themselves on outside walls or on garden plants. Using special bunches of hairs under their feet and a safety line in case of mishap, they have no trouble at all walking up vertical walls and often come inside houses – hence the name. They are found in the North Island and as far south as Christchurch, where they often build a silken retreat in the cracks in walls. When the sun goes behind a cloud, they will quickly vanish inside. They will frequently leap onto a fly from 5 cm away. Like many jumping spiders, the males and females are often seen performing courtship dances. The Australian origin of some became obvious only recently when it was realised that the spiders found here were in fact two slightly different species.

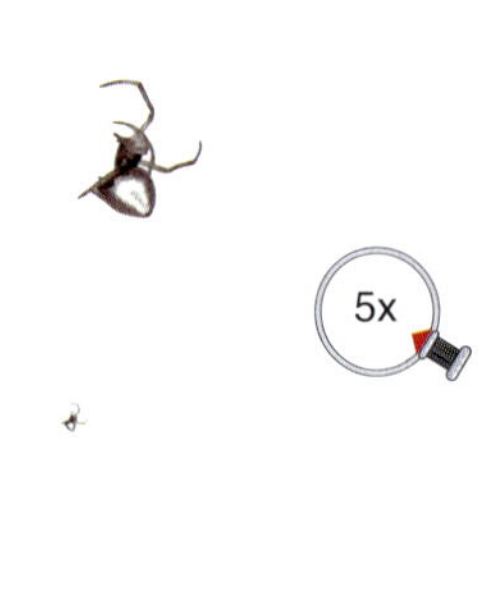

Various colour schemes of **house hopper spider**. The picture on the left shows the typical colouring of the male.

Bronze Aussie Jumper

Helpis minitabunda

This east Australian spider was first noticed in Auckland in 1972, but is now common in the North Island at least as far south as Palmerston North, where it is found in tall grasses and leafy shrubs. Also common inside houses. The spider's common name – although not well known here – is the one by which it is known in its Australian homeland.

DAY HUNTERS

Where no magnification is indicated, photos are shown life-size.

A typical flat, silken, jumping spider retreat (left) made in this case on the underside of a leaf of New Zealand flax by a **bronze Aussie jumper** (right) – for more details, refer to previous page.

Cream Jumping Spider

[An apparently undescribed species]

Native Striking for its pale colouring, this tiny jumping spider lives in native scrub. Since no arachnologist (spider expert) has been able to identify it and because of the native habitat in which it was found, it is likely to be a 'new' native species. Photographed in January 2005, high above Tairua on the Coromandel.

Polkadot Hopper Spider

Opisthoncus polyphemus

Australian. From coastal New South Wales and Queensland; also Papua New Guinea. Has distinctive polka-dot markings on its head. Photographed in September 2000 on a Poor Knights lily in an estuary-side garden at Mangawhai, Northland. Not previously recorded in New Zealand.

Goldenbrown Jumping Spider

Trite auricoma

Native When stalking its prey, this common spider pounces from about 3–5 cm away. It doesn't jump as readily as some of the other species, so is sometimes called a 'runner' rather than a 'jumper'. It is common in dry, rolled-up flax leaves or cabbage tree leaves, under stones, in shrubs, low vegetation and on the ground. Both examples shown here were photographed in tall native forest along the ridgeline of the Coromandel Range. A clear horizontal band of pale yellow hairs below the eyes of the spider on the right identifies it clearly as a mature male.

Flattened Jumping Spider

Holoplatys species

Native Members of this group of jumping spiders (which are also known as **crevice jumping spiders** or **chink jumping spiders**, all *Holoplatys* species) all look as if they have been squashed flat – like a flower pressed between the pages of a book. This curious shape allows them to slide into narrow cracks and crevices to hide, retreating for example beneath the bark of trees or between the overlapping boards of a fence, or – like the one illustrated below – vanishing into a loose-fitting timber joint of a gate in Christchurch. Tucked away in such unlikely hideaways, they seem to leap out of nowhere to catch their passing prey.

A **flattened jumping spider** catching a blowfly.

Lynx Spiders

[Family: Oxyopidae]

These spiders are believed to have particularly good vision, which is why they were named after the keen-eyed wild cat, the lynx. During the day, they are often seen resting among the leaves of plants from where they leap out and grab small insects. They also hunt actively, running and leaping after their prey. One species from Florida (*Peucetia* species) even spits venom at its enemies; this spray can reach 20 cm and stings if you get it in your eyes. Thankfully, though, our New Zealand species is not a spitter. Lynx spiders have eight eyes arranged in three rows, like a hexagon: two, four, two. Most of them have pointed bodies. Over 400 species are known worldwide, of which only one named species is known in New Zealand.

Lynx Spider

Oxyopes gracilipes (was *Oxyopes gregarius*)

Native Straw-coloured. Common throughout the North Island and parts of the South Island. It is often seen resting on flowers and leaves in bright summer sunshine but also frequently hunts more actively, running and hopping around, capturing small insects on low shrubs and hedges. It is found not only in gardens, but also in native scrubland and along forest edges. This one is enjoying the sunshine on a lavender flower. In spring or summer, the female lays a flattened egg sac, usually on a leaf. She stands guard at first, straddling this to protect it from hungry predators.

Fleet-Footed Spiders

[Family: Corinnidae]

As the common name of these spiders suggests, fleet-footed spiders are fast runners. Members of this family are also appropriately known as **sun spiders**, because they hunt mostly during the day, often in bright sunshine. Most members of the family mimic ants – something which distinguishes them from members of the family Clubionidae, or **sac spiders**, with which they were previously included. Over 900 species are known worldwide, of which only one (shown below) is known in New Zealand.

6x

Australian Ground Spider

Nyssus coloripes (was *Supunna picta*)

From Australia. Common in homes and gardens, on roads, dry open country and riverbeds. It is easy to recognise simply by the crazy way it moves. It seems to dash all over the place, often in full sun, at amazing speed, looking more like a clockwork toy out of control than a spider, as it holds its two orange front legs up, waving them about like the 'feelers' of a hunting wasp. Like other members of its family, it builds a silken sac to hide in. In its native Australia it is appropriately known by the far more descriptive names of **wasp-mimicking sac spider** or **spotted ground swift spider**.

DAY HUNTERS

Wolf Spiders

[Family: Lycosidae]

Called wolf spiders because of the way they run down their prey like wild dogs or wolves. Some of these ground-dwellers hunt during the day and some at night, skilfully finding their way back home by noting the angle of the polarised light around them. Their sensitive eyes also glow at night, especially if you shine a torch directly at them. They have eight eyes in three rows. Two large eyes, one on each side of the top of their head, can be used for looking behind. In front, two more large eyes point forward, with a further four tiny eyes in a row underneath. For insect control, wolf spiders are among the most useful groups of spiders, many individuals consuming about 5–15 small insects a day. Some rice growers overseas even go to the trouble of leaving a wild strip around their fields for the spiders to live and breed in, to encourage them to patrol their fields for leafhopper insects. The female wolf spider drags her egg sac around with her, then later carries her young on her back. She is equipped with special 'foothold hairs' for her babies to hang on to.

28 species now known in New Zealand; over 2200 worldwide.

3x

Garden Wolf Spider

Anoteropsis hilaris (was *Lycosa hilaris*)

Native Commonly seen on sunny summer days running around in pasture and gardens. For five weeks, the female carries her white, ball-shaped egg sac around with her, tucked underneath her body. When her 100 or so young hatch out, they climb onto her back to be carried around for another week, growing until they are old enough to go off independently and look after themselves. Found not only in gardens but also in its original habitat, native herbfields and tussock country, so is also known as the **striped wolf spider**.

The **garden wolf spider** piggybacking her babies. (Three have just slipped off.)

Where no magnification is indicated, photos are shown life-size.

Seashore Wolf Spider

Anoteropsis litoralis

Native Very hard to spot until it makes a dash. The instant it freezes, it seems to disappear. This is one of several, similar-looking, speckled grey spiders which live on beaches and sand dunes. This species is seen around the North Island and Canterbury, and can be different colours depending on the sand in which it lives. The adults are more active at night; young ones hunt during the day.

Bush Wolf Spider

Allotrochosina schauinslandi

Native Lives in damp forest throughout New Zealand, where it is often found hiding during the day among rotting logs or forest leaf litter. Although also sometimes seen under boards or clumps of grass, it cannot stand dry conditions. Unlike the **garden wolf spider** (opposite) it tends to roam and feed only at night.

Prowling Spiders

[Families: Miturgidae & Zoropsidae]

Most of these hunting spiders are active only at night, when they prowl the forest floor. During the day, they prefer to hide under logs or among leaf litter to help protect them from being hunted by large spider wasps. Prowling spiders belong to a wider group which build silken sacs to hide in, so are sometimes collectively known as **sac spiders** or (to distinguish them from members of the Clubionidae family) as **large sac spiders**. Over 400 species of Miturgidae and Zoropsidae are known worldwide, of which more than 40 are known in New Zealand – all these large-bodied spiders are found throughout the country.

Two **females** – which tend to look chunkier than the males (opposite page).

3x

Where no magnification is indicated, photos are shown life-size.

The **male prowling spiders** are more slender, with longer legs and a telltale splash of salmon pink on their palps. When mature, these long-legged 'boys' roam for a mate, so are more often seen, especially when they come into houses by mistake.

Large Brown Vagrant Spider

Uliodon species [Family: Zoropsidae]
(was *Miturga* species [Family: Miturgidae])

Native Found throughout New Zealand, mostly in forest under logs and rocks from where they emerge at night to hunt on the forest floor. By day, they usually remain well protected inside a silken sac. When necessary, they can run surprisingly fast, then suddenly freeze before running again. At least 40 species are now known in New Zealand.

These large palps ('feelers') are a typical feature of male spiders.

2x

NIGHT HUNTERS

Where no magnification is indicated, photos are shown life-size.

Whitetailed Spiders

[Family: Lamponidae]

Previously grouped along with **vagabond spiders** [Family: Gnaphosidae]. 'Whitetails' hunt mostly in the evening and at night, hiding in a silken sac during the day. Indeed, as part of a wider group of families [Clubionidae, Miturgidae and Gnaphosidae], they are also known as **sac spiders**. Two very similar introduced species of Lamponidae are found in New Zealand; about 200 species worldwide.

These extra markings are seen on juvenile whitetails.

Adult whitetail, showing the white tip which gives the spider its name.

Where no magnification is indicated, photos are shown life-size.

Whitetailed Spider

Lampona species

Australian. Two similar-looking species (*Lampona cylindrata* and *L. murina*) live in New Zealand. Indeed, to distinguish them, experts need a microscope. Both have a narrow abdomen shaped like a lemon pip, with a whitish tip. Adults are 1–2.5 cm long; the smaller young have extra white markings on the sides. Unusual in that they eat only other spiders, 'whitetails' often pluck at another spider's web (like that of a **grey house spider**), pretending to be a captured insect. When the web's owner appears, it is grabbed and eaten (see below).

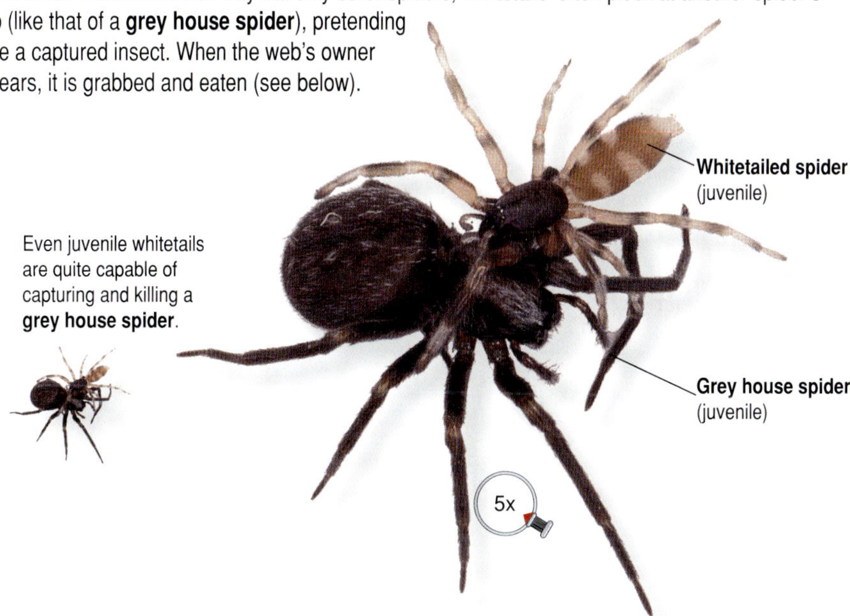

Even juvenile whitetails are quite capable of capturing and killing a **grey house spider**.

Whitetails are very common in gardens and houses, but are generally active only at night, particularly from spring to late autumn. Although not aggressive towards humans, they will bite if provoked.

Although first noticed here in the 1800s, it was not until 1980 that New Zealanders began blaming the bite of the whitetail for causing necrotic ulcers (ulcers which are slow to heal). Dramatic press headlines followed: 'Fears of Biting Spider Plague', 'Spider Suspect in Death Mystery'. However, tests of the venom have never shown a significant poison likely to cause ulcers. The alarming newspaper reports continued . . .

Then, in 1989, Adelaide Children's Hospital ran a trial, collecting information on 36 cases of spider bites where the spider could be identified, a survey which was repeated by Christchurch Hospital (2001–2003) on people in New Zealand believed to have had contact with venomous spiders. In neither study could a single instance be found of whitetailed spider bites causing necrotic ulcers.

Another comprehensive study (published in 2003) covered data on 130 definite bites by *Lampona cylindrata* or *L. murina* reported over a three year period to Australian poison information centres and emergency departments. In each case, the spider was identified and, in most cases, the offending spider had been trapped against the person's skin. The bite was almost always painful at the time (like a definite pinprick), but never with any serious effects.

So, while it is certainly a good idea to avoid being bitten by any spider, it has to be said that there is no evidence that whitetails are especially dangerous. (For further background notes on whitetail bites, see page 60.)

Six-Eyed Spiders

[Family: Dysderidae]

During the day, these spiders hide in a silken retreat among stones, or in holes in bark or wood. Generally, they come out only at night to feed. Many species lay down trip-up threads from this safe retreat to trap passing prey. As their common name suggests, most of the spiders in this group have six eyes (rather than the usual eight) and these are arranged more or less in a circle. (The same is true of several other families of New Zealand spiders: the Orsolobidae, Periegopidae, Segestriidae, Scytodidae and Oonopidae.) These spiders are renowned for their particularly long, sharp fangs – although they rarely bite people. Worldwide, about 500 species are known, most of them native to countries around the Mediterranean. Only one has so far found its way to New Zealand.

Slater Spider

Dysdera crocata

Introduced. Widespread and common in compost heaps, beneath bricks and rocks; also in cool, damp houses, from where it comes out at night to feed. It eats slaters, so can be particularly useful in the garden. (Some people don't like slaters.) Its back end looks remarkably like a roasted peanut. This spider is found in Europe too where it is known as the **woodlouse spider** ('woodlouse' being the English name for 'slater'.) It is best not to trap one against your skin, for it can bite, causing a painful swelling which can be slow to heal. Its tightly-woven silken retreat serves a double purpose: if it gets flooded, this holds a significant reserve of air which the submerged spider can breathe. This supply can last ten times longer than air held without a sac. Also known as the **six-eyed garden spider**.

Back end looks like a roasted peanut.

Slaters are a popular food of this common garden spider.

Where no magnification is indicated, photos are shown life-size.

Huntsman Spiders

[Family: Sparassidae (was Heteropodidae)]

Because of their flattened body-shape and crab-like legs, these magnificent creatures are also known as **giant crab spiders**. They weave no web and hunt at night. The family includes the **tarantulas** of Australia and the large **banana spider** (*Heteropoda venatoria*) which is occasionally brought into New Zealand accidentally with bananas. The worldwide figure for the number of known species is about 1000, only one of which is established in New Zealand.

Avondale Spider

Delena cancerides

Australian. This spectacular-looking spider arrived in the Avondale district of Auckland in the early 1920s in a packet of Australian timber, and hasn't spread much since. It is most unusual in that it lives year-round in large communities of up to 300, mostly under loose bark of wattle trees, although males do occasionally come into houses too. Hunts and feeds at night; the young working in co-operative teams. Body can be up to 3 cm long, legspan 20 cm (the size of a saucer). They are not at all aggressive and are relatively harmless, making them ideal for film-making. Indeed, 374 were sent from New Zealand to Hollywood in 1989 for the film, *Arachnophobia*. Known overseas as the **South Australian huntsman**.

Nurseryweb & Water Spiders

[Family: Pisauridae]

Large, long-legged spiders, most of which build a nurseryweb for their hatchling young. They don't use these webs for catching food, but instead chase their prey – usually at night. Though large, they are harmless to people. With the recent discovery of unnamed **water spiders** (*Dolomedes* species), New Zealand is now known to have at least three species. Worldwide, the figure is over 300. If you notice a likeness between the patterning of these spiders and the markings of **wolf spiders**, you are not the first; both belong to the same group [Superfamily: Lycosoidae], however **nurseryweb spiders** have smaller eyes.

Left, a **water spider** (*Dolomedes* species) spreading itself out like a cloak on the surface of a slow-flowing stream.

Below, the empty skin (moult) of a **water spider**. In summer, all spiders, when they are actively growing, will eventually find their skin is getting too tight for them so must climb out of it. The spider's soft body beneath it then hardens to form a new skin.

Water Spider

Dolomedes aquaticus

Native Can walk or run on top of water, even swim or dive sometimes – indeed, they have been known to stay underwater for up to half an hour. With legs spread, it is often seen resting on the water surface, but can also float downstream, jumping if need be across rougher patches of swift current. It hunts along the edges of forest streams, shingle river-beds and lake shores – mostly at night. In its usual pose, it rests its front two legs on the water surface, waiting for vibrations of struggling insects and even small passing fish. A recently-discovered sister species of water spider (also *Dolomedes*) has been observed building nurserywebs for its young, but *Dolomedes aquaticus* itself builds no nurseryweb. There are now known to be several species of water spider in New Zealand. The one illustrated above (photographed in a forest stream near Waikawau Bay on the Coromandel Peninsula in March, 1998) is believed – because of the striking patterning of its body – to be an undescribed species.

Where no magnification is indicated, photos are shown life-size.

The immature **nurseryweb spider** is paler.

Nurseryweb Spider

Dolomedes minor

Native Although the nurserywebs themselves are often noticed, particularly in summer, on the tips of branches of low shrubs like gorse, hakea, broom and mānuka, the spider itself is rarely seen. These nurserywebs are not used for catching prey. Instead, this spider wanders for its food – mostly at night. In late spring or early summer, the female carries a huge egg sac. After about five weeks – just as her 200 or so spiderlings start to hatch inside the sac – she attaches this sac near the top of a bush and proceeds to build a protective nurseryweb around it. Come out with a torch at night to find her straddled on this nurseryweb (photo bottom left), guarding her babies. By day she is usually hiding in the undergrowth nearby. After her young have spent a week growing inside this cosy little nursery, they climb out; before long they will be letting out a long silk thread to go 'ballooning' (see page 4).

Adult female **nurseryweb spider** guarding her nurseryweb.

Australian golden hunting wasp (*Cryptocheilus australis*) dragging a paralysed **nurseryweb spider** back to her nest as food for her young.

Four-Lunged Spiders

[Family: Gradungulidae]

These spiders are sometimes called 'living fossils' – an arachnid version of the tuatara. They are half like modern spiders (with fangs that nip together from the sides) and half like ancient spiders (with four book-like lungs instead of two). In New Zealand, three species have been named; a further 12 live in Australia.

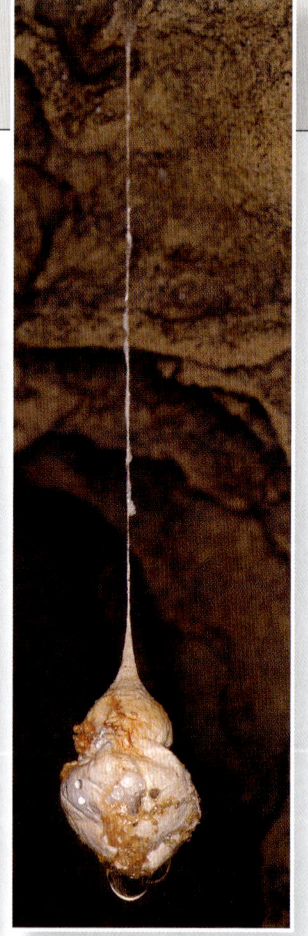

The spider's egg sac hanging from the cave roof.

Nelson Cave Spider

Spelungula cavernicola

Native New Zealand's largest native spider, with a leg span of up to 15 cm (the width of this page). It is so rare that it remained undiscovered until 1957 and is now a protected species. The white ball dangling like a Christmas decoration on a long thread is its egg sac. Eighty to ninety spiders will soon hatch out of this and start eating many of their own brothers and sisters. A few weeks later, just six or so will be left to grow into adults. As adults, they eat cave wētā. This one was photographed in a forest cave near Karamea in the South Island.

Where no magnification is indicated, photos are shown life-size.

OTHER ARACHNIDS

(Eight-legged invertebrates with no waist)

Unlike spiders, none of the following creatures have a narrow waist or poison fangs. Although some can make webs to protect their young, none can make any kind of food-catching web. (Includes **harvestmen**, **mites**, **ticks**, **false scorpions** and **sea spiders**.)

False Scorpions

[Order: Pseudoscorpiones (=Chelonethi)]

Apart from the absence of a stinging tail, these little creatures look like tiny scorpions – hence the name. Their bodies are flat, usually brown or pinkish. Several native species are found in forest, under the bark of dead trees and in leaf litter; others live on the seashore or in the mountains. They eat mainly springtails (and also beetle larvae) which they subdue with poison from their pincers (not from fangs). This poison is not sufficient to harm people. The mother protects her young in a pouch beneath her body – rather like a kangaroo. So far, 70 species are known in New Zealand (most of them native) and about 3300 worldwide.

Life cycle: egg > larva > nymph > adult false scorpion

Fortunately, the poison pincers of **false scorpions** are harmless to people.

False Scorpion

[Order: Pseudoscorpiones]

Native A typical example of the tiny species commonly found in New Zealand forest leaf litter. Fortunately, although these creatures look rather like the true stinging scorpions, they are completely harmless to people.

Harvestmen

[Order: Opiliones]

The common name of these arachnids comes from the European one below, which is often noticed in the fields of Europe at harvest-time. Unlike spiders, harvestmen have oval-shaped bodies with no waist. They lack poison fangs and cannot spin silk. Those with eyes have only two perched on top of a little bump on their head – like a submarine conning tower. At night, they eat small insects (either living or dead). Many can protect themselves by exuding a foul-smelling fluid from the front of their bodies. They fall neatly into three distinct groups: the **longlegged harvestmen** [Suborder: Palpatores], the **shortlegged harvestmen** [Suborder: Laniatores]; and the tiny **mite-like harvestmen** [Suborder: Cyphophthalmi]. New Zealand has over 160 native species, most of them living among the leaf litter in native forest. The number of species known worldwide is about 5500.

Life cycle: eggs hatch as little harvestmen

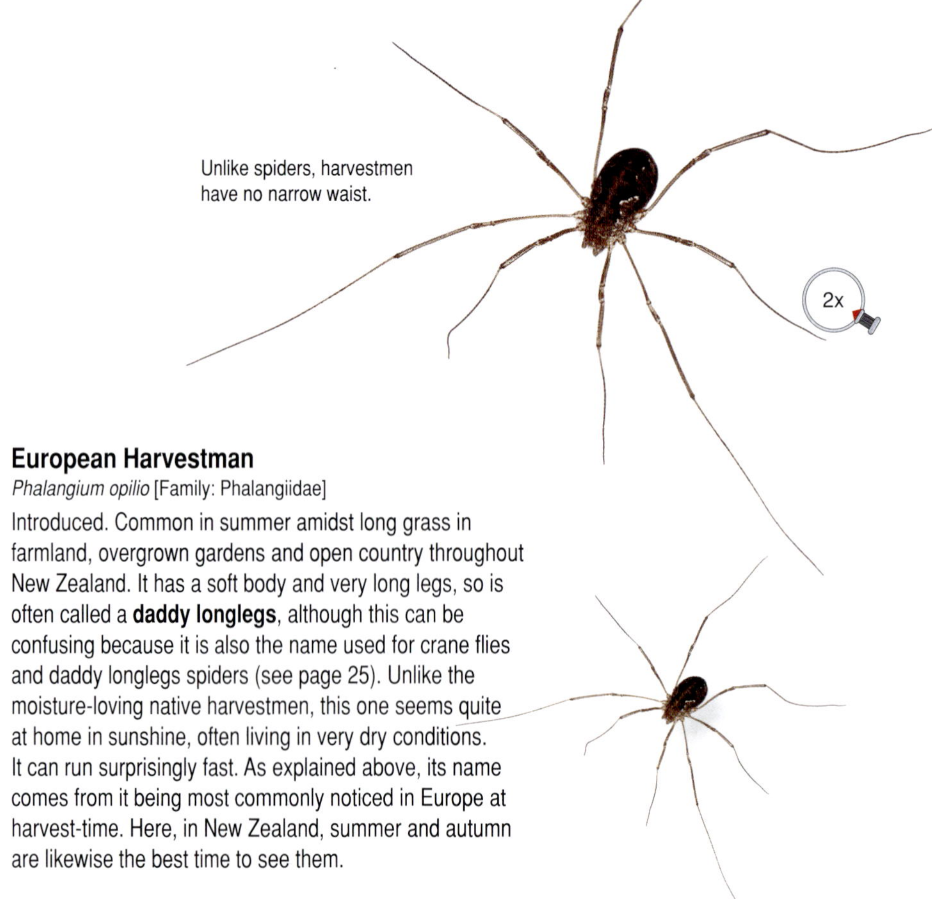

Unlike spiders, harvestmen have no narrow waist.

European Harvestman

Phalangium opilio [Family: Phalangiidae]

Introduced. Common in summer amidst long grass in farmland, overgrown gardens and open country throughout New Zealand. It has a soft body and very long legs, so is often called a **daddy longlegs**, although this can be confusing because it is also the name used for crane flies and daddy longlegs spiders (see page 25). Unlike the moisture-loving native harvestmen, this one seems quite at home in sunshine, often living in very dry conditions. It can run surprisingly fast. As explained above, its name comes from it being most commonly noticed in Europe at harvest-time. Here, in New Zealand, summer and autumn are likewise the best time to see them.

Where no magnification is indicated, photos are shown life-size.

Longlegged Harvestman

Megalopsalis species [Family: Monoscutidae]

Native Has long, bendy, thread-like legs which break off easily – a feature which can be useful to these creatures when escaping a predator. The males and females look so different from one another that they were at one time thought to be different species; the female is colourfully patterned while the male (pictured above) is generally black. There are many such native species in New Zealand but few have been studied or named. This one was photographed inside a South Island cave, where it was living among cave wētā and cave spiders.

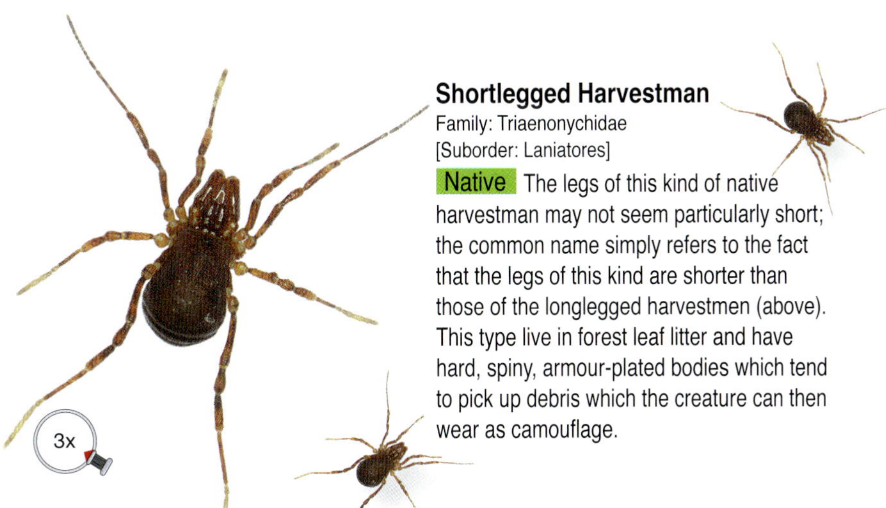

Shortlegged Harvestman

Family: Triaenonychidae
[Suborder: Laniatores]

Native The legs of this kind of native harvestman may not seem particularly short; the common name simply refers to the fact that the legs of this kind are shorter than those of the longlegged harvestmen (above). This type live in forest leaf litter and have hard, spiny, armour-plated bodies which tend to pick up debris which the creature can then wear as camouflage.

Mites

[Subclass: Acari]

Most of these biting, sucking creatures are less than 1 mm long, although some can be up to 3 mm. Some are free-living; others live on specific plants or animals. A few are a nuisance in orchards, in gardens or on animals; examples include the **scabies mite** (*Sarcoptes scabiei*) on human skin, **housedust mite** (*Dermatophagoides* species) in carpets and **varroa mite** (*Varroa destructor*) on honey bees. Others are helpful. Many native species are found among forest leaf litter, where some eat other creatures, while others feed on plants or fungi. The adults have eight legs (as one would expect), but their young (larvae) have only four or six. New Zealand has several thousand species (about 840 of them native). Over 30,000 worldwide.

Life cycle: egg > larva > nymph > adult mite

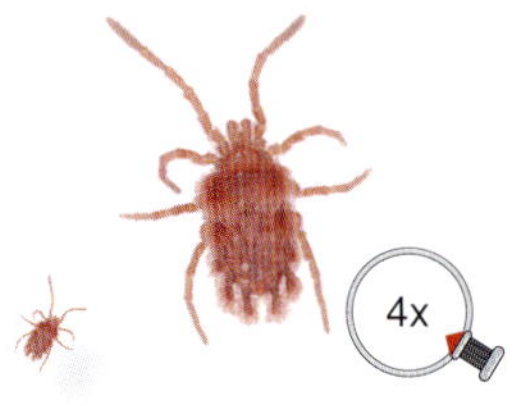

Large Hairy Bark Mite

Cheyzeria species [Family: Cheyzeridae]

Native One of New Zealand's larger mites. A common predator under bark and stones. The large tufts of hairs on its body are unusual for a mite.

Predatory Whirligig Mite

Anystis species [Family: Anystidae]

Introduced. A very common bright red mite seen crawling up walls and trunks of trees. Useful in orchards, because it eats harmful spider mites which damage the fruit and leaves.

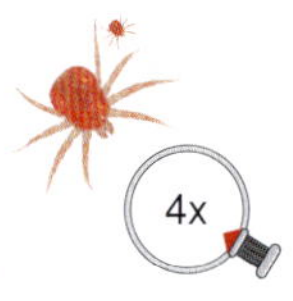

Lacebark Gall Mite

Eriophyes hoheriae [Family: Eriophyidae]

Native As food for its young, this mite makes a stem gall on the branchlets of native lacebark trees (*Hoheria populnea* and *H. sexstylosa*).

Where no magnification is indicated, photos are shown life-size.

Gorse bush covered with protective webbing made by **gorse spider mites**. (Not life-size)

Gorse Spider Mite

Tetranychus lintearius [Family: Tetranychidae]

First imported from England for gorse control in 1988 and now established in all parts of New Zealand. The spiky gorse leaves are sucked by the mite, turn whitish or brown, and the flowers drop off. Although this rarely kills the plant, the mite is nevertheless a useful tool in the effective control of gorse. (Avoid using pesticides on mite-infested plants.) Found all year – only on gorse bushes – but easier to find in the warmer months. Their webbing is often destroyed by rain.

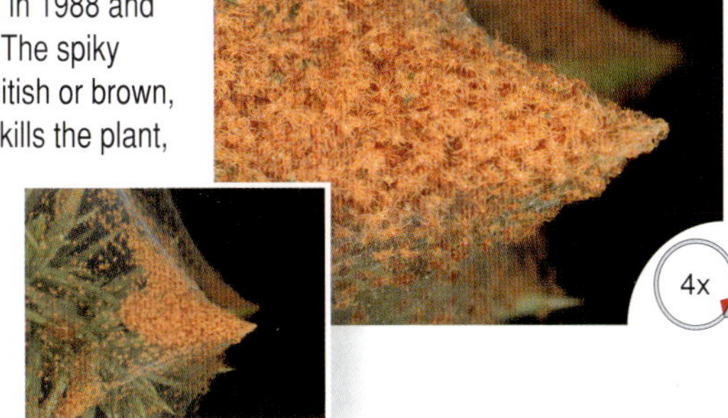

Ticks

[Subclass: Acari, Order: Ixodida]

Like their smaller cousins, the **mites**, ticks are biting, sucking creatures. They feed on the blood of birds, mammals and reptiles and fall into two main groups: the **hard ticks** [Families: Nuttalliellidae and Ixodidae] and the **soft ticks** [Family: Argasidae]. Hard ticks are flat-bodied creatures (whose body swells up when they feed). The soft ones are usually round and berry-like. Both kinds are responsible for spreading certain diseases. Some ticks have eyes; some don't. Although the adults always have eight legs, their young (larvae) have only four or six. Over 800 species are known worldwide; but only six native species so far known in New Zealand, each with its own host (neatly described by its common name) – the **bat tick**, **tuatara tick**, **kiwi tick**, **cormorant tick**, **penguin tick** and **common seabird tick**. New Zealand's best known example, however, is an introduced species, the so-called **New Zealand cattle tick** (illustrated below).

Life cycle: egg > larva > nymph > adult tick

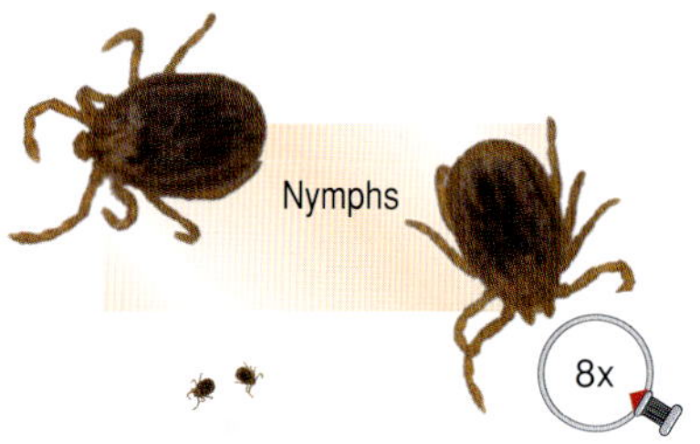

New Zealand Cattle Tick

Haemaphysalis longicornis [Family: Ixodidae]

Introduced. In spite of its common name, New Zealand's most commonly noticed tick comes from Asia. Besides being found on cows here, it is also common in Australia and on some of the Pacific Islands. The adult females, nymphs and six-legged larvae all bite cattle and other large animals (including dogs and people) to suck their blood. Although this tick is found as far south as Golden Bay in the South Island, it is a particular problem for farmers in the northern half of the North Island. Remarkably, the female tick requires no male to mate. Indeed, if male cattle ticks exist at all in New Zealand, they are very rare. In early summer, the female lays 1000–1500 eggs in the grass. These hatch into six-legged larvae which climb up the plants to wait for a passing animal. When their chance comes, they hop onto the animal where they stay to feed.

Where no magnification is indicated, photos are shown life-size.

Sea Spiders

[Class: Pycnogonida]

Most sea spiders have incredibly slender bodies and four pairs of walking legs (although a few do have five or six pairs). Unique to sea spiders is an extra pair of legs which the female uses to transfer her eggs to the male. This might seem strange enough but the male then glues these eggs onto his own front egg-legs. Most coastal species are only a few millimetres across, but the legspans of some deep sea species can extend for up to half a metre. In the Antarctic, large species are common even near the shore. It has long been a puzzle where sea spiders fit in the scheme of things, but they are now known to be closely related to mites and ticks. Including species which have been collected from around the Snares Islands, 23 species are known in New Zealand waters, many of them living in water deeper than a thousand metres. The worldwide total is likely to be well over 400 species.

Life cycle: egg > nymph > adult sea spider

4x

Coastal Sea Spider

[Class: Pycnogonida]

Native The sea spiders in this picture are common between high and low tide, but can be hard to find because of their small size and colouring. Coastal species like this one eat sea anemones, sea slugs, corals, and other strange marine creatures such as moss animals (Bryozoans), hydroids and bristleworms (polychaete worms). The four shown in the photo are frantically wriggling, apparently struggling to hide from the light. They were found living inside a shipworm hole inside a piece of driftwood washed up on the shores of Auckland.

Recommended Reading

Spiders of New Zealand and their Worldwide Kin by Ray & Lyn Forster. University of Otago Press, Dunedin, 1999.

Fascinating Spiders by Olwyn Green and Mavis Lessiter. The Bush Press, Auckland, 1987.

The Book of the Spider: From Arachnophobia to The Love of Spiders by Paul Hillyard. Hutchinson, London, 1994.

'Eight legs, two fangs and an attitude' by David Faulls. *New Zealand Geographic*, No. 10, 1991: 68–96.

Spiders in New Zealand by Bill Fairweather. Penguin, Auckland, 2008.

Photographic Guide to Spiders of New Zealand by Cor J. Vink. New Holland, Auckland, 2015.

Background Notes

More about Whitetailed Spider bites

Anyone who has ever suffered from a necrotic ulcer, or who has been shown the scar of someone who has, is likely to remain unconvinced by reassurances about the harmless nature of whitetailed spider bites. In fact, in many cases, the victim's doctor will confirm that a whitetailed spider bite was the most probable cause of the ulcer.

For almost a century in New Zealand (prior to 1980), this spider was not blamed in this way and went more or less unnoticed. It is true, however, that one of our two whitetailed spider species (*Lampona cylindrata*) became more common around that time – but only in the South Island.

Researchers investigating this first analysed the whitetail venom, looking for any kind of poison in the spider's venom glands likely to cause ulcers. Finding nothing especially poisonous, they wondered whether the spider was causing problems by other means. Was it perhaps picking up a soil-dwelling microbacterium on its fangs, for example?

Meanwhile, since the suspect was an Australian, researchers were looking on both sides of the Tasman for patterns in the records of verifiable spider bites. Since necrotic ulcers can be triggered by any puncture of the skin, they had to consider whether the whitetail was being mistakenly blamed. Perhaps doctors had joined everyone else in pointing the finger at an innocent spider, simply because of its bad reputation?

The detail of all this research and the results the researchers found are impressive. Throughout Australia and New Zealand, they could not find a single case of a person who had developed a necrotic ulcer who had seen and caught a whitetailed spider in the act of biting (which is odd because the spider has poor eyesight and is easily caught). Many people claiming to have been bitten by a whitetailed spider had to admit that they saw no spider, saying that they must have been bitten while asleep. This is again odd since the bite of a whitetail is known to be painful (like the bite of most spiders). Anyone bitten by one while they are sleeping is likely to wake up immediately.

On the other hand, in every case where a whitetailed spider was indeed caught in the act of biting, no ulcers from the bite resulted.

So from all that is now known, it has to be said that there is no evidence that whitetails are especially dangerous. If you do see and feel a spider biting you, use a container to catch it, then clean and check the wound. Like any other wound, keep a spider bite clean and covered. In the event of an adverse reaction seek medical help.

For further reliable information on the whitetailed spider (and to keep up to date with new research on the subject), refer to any of the following official websites:

www.health.govt.nz	Ministry of Health
www.landcareresearch.co.nz	Landcare Research
www.tepapa.govt.nz	Te Papa Museum, Wellington

For health professionals, the relevant *Medical Journal* articles are:

'White-tail spider bite: a prospective study of 130 definite bites by Lampona species'. Geoffrey K. Isbister and Michael R. Gray. *Medical Journal of Australia*, 18 August 2003; 179: 199–202. (Available online at www.mja.com.au)

'White-tailed spider bites – arachnophobic fallout?' Banks, J., Sirvid, P. and Vink C. *New Zealand Medical Journal*, 2004, 30 January. Vol. 117; no. 1188. (Available online at www.nzma.org.nz/journal)

More about the Daddy Longlegs Spider

Another popular myth going around at the moment concerns the familiar daddy longlegs spider having an especially poisonous venom (refer to page 25). Recently, this tale has been further embellished with claims about the whitetailed spider, saying that by eating these long-legged spiders, the whitetail is able to accumulate the toxic venom of daddy longlegs spiders to pass on to humans. This makes a good story, perhaps, but it certainly isn't true!

Glossary

Where possible, this book avoids using technical terms (or explains them in the text). However, for those who wish to know the scientific terms used for the various parts of a spider, there is a labelled drawing of all their bits on page 2.

Index (Where several page numbers are given, the main entry is shown in **bold** type.)

About the Author

Andrew Crowe continues to share his love of nature through his books, finding innovative ways of presenting natural history to all ages in an easy-to-follow style. He conceives of, researches and designs his books himself. With over 40 to his name, he has been shortlisted 31 times for various New Zealand book awards. Six of his titles have won major awards, including the non-fiction award at the NZ Post Children's Book Awards and a Storylines Notable Book Award for ***Which New Zealand Spider?***. In 2009, he received the Margaret Mahy Award for his services to children's literature.

Other Andrew Crowe guides in this ***Which?*** series:

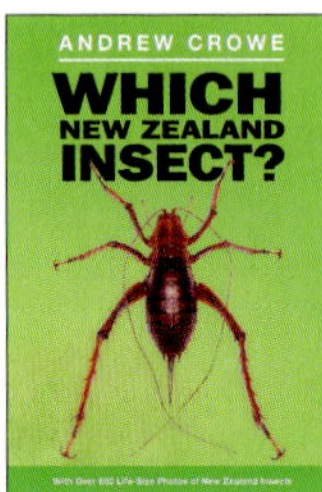

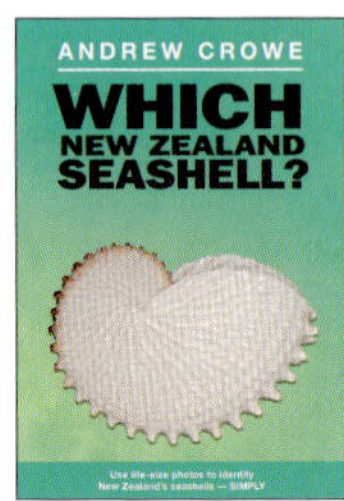

New Zealand creatures

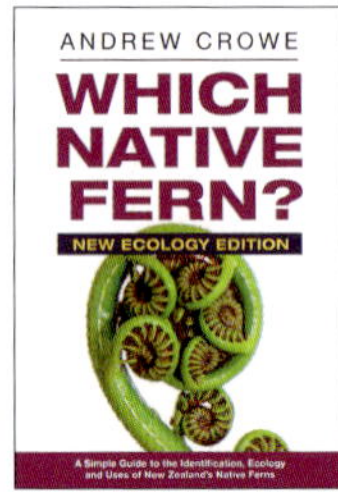

New Zealand plants

Andrew Crowe has also produced laminated, all-weather fold-out guides in the ***Flip-Guide*** series, a series of pocket-sized ***Mini Guides*** and a range of very popular large-format family books in the ***Life-Size*** series, all published by Penguin Books.

For more about the author, see:
https://authors.org.nz/author/andrewcrowe/